YOUR KNOWLEDGE HAS VALUE

John Bredakis

A unique approach to Fourier's series and integral

With complete proofs for all the topics discussed

GRIN Verlag

Bibliografische Information der Deutschen Nationalbibliothek:

Die Deutsche Bibliothek verzeichnet diese Publikation in der Deutschen National-
bibliografie; detaillierte bibliografische Daten sind im Internet über http://dnb.d-
nb.de/ abrufbar.

Imprint:

Copyright © 2011 GRIN Verlag GmbH
Druck und Bindung: Books on Demand GmbH, Norderstedt Germany
ISBN: 978-3-640-89690-5

This book at GRIN:

http://www.grin.com/en/e-book/170707/a-unique-approach-to-fourier-s-series-and-
integral

GRIN - Your knowledge has value

Der GRIN Verlag publiziert seit 1998 wissenschaftliche Arbeiten von Studenten,
Hochschullehrern und anderen Akademikern als eBook und gedrucktes Buch. Die
Verlagswebsite www.grin.com ist die ideale Plattform zur Veröffentlichung von
Hausarbeiten, Abschlussarbeiten, wissenschaftlichen Aufsätzen, Dissertationen
und Fachbüchern.

Visit us on the internet:

http://www.grin.com/

http://www.facebook.com/grincom

http://www.twitter.com/grin_com

John Bredakis

A unique approach to Fourier's series and integral

With complete proofs
for all the topics discussed

Essay

G R I N

Verlag fur akademische Texts

A unique approach to Fourier's series and integral

$$f(x) = Sf(x) + rn(x) \quad \square \quad Sf(x) = [ao] + \sum_{k=1}^{+\infty} ak.\cos(k.x) + \sum_{k=1}^{+\infty} bk.\sin(k.x)$$

Sf(x)=Trigonometric serial expansion of the function f(x)
rn(x)=Remainder function
Fourier's coefficients [ao],ak,bk k=1,2,3,...,n (n->+oo)

The appropriate selected f(x)   rn(x)->0

1.   The function f(x) is continuous in (-π,0) and in (0,π)

2. At x=0 this f(x) is either continuous or piecewise continuous

3. The function f(x) is considered as periodically continued on
   the whole real axis x (from -oo to +oo) with period 2π.

4.   By periodicity: f(π+)=f(-π+) and f(-π-)=f(π-)

    Or the f(x) is a smooth curve over the semiperiod (0,π)
    This f(x) is considered odd or even function in (-π,π).
    And the 2. 3. and 4. of the above.

.An odd function f(t) by an even function cos(k.t) makes an odd
function f(t).cos(k.t),which drops out in [-π,π].

.An even function f(t) by an odd function sin(k.t) makes an odd
function f(t).sin(k.t),which drops out in [-π,π].

t is the variable , while x is fixed (-π< x <π)

$$\cos[k.(t-x)] = \cos(k.x).\cos(k.t) + \sin(k.x).\sin(k.t)$$

$$Sf(x) = \frac{1}{\pi} . \int_{-\pi}^{\pi} f(t).\left[\frac{1}{2} + \sum_{k=1}^{+\infty} \cos[k.(t-x)]\right].dt = \frac{f(x-)+f(x+)}{2} = f(x)$$

* Iff f(x) is continuous at selected fixed x

Fourier,watching the tidal waves,observed that waves,which are
functions of sine and cosine , colloiding with each other were
converted into curves , which are functions of x.

Ιωάννης.Κ.Μπρεδάκης   **John.K.Bredakis MD**
American Board Certified Cardiologist   Athens Greece 2011

By orthogonality properties of trigonometric intergrals

in the given interval of integration

$$\frac{1}{\pi} \cdot \int_{-\pi}^{\pi} \left[\frac{1}{2} + \sum_{k=1}^{+\infty} \cos[k.(t-x)] \right].dt = \boxed{1} = \frac{1}{2} + \frac{1}{2}$$

t is the variable , while x is fixed $(-\pi < x < \pi)$

$$\cos[k.(t-x)] = \cos(k.x).\cos(k.t) + \sin(k.x).\sin(k.t)$$

$$Sf(x) = \frac{1}{\pi} \cdot \int_{-\pi}^{\pi} f(t). \left[\frac{1}{2} + \sum_{k=1}^{+\infty} \cos[k.(t-x)] \right].dt = \frac{f(x-)+f(x+)}{2} \ ^* = f(x)$$

* Iff f(x) is continuous at selected fixed x

Orthogonality properties of trigonometric integrals
in the domain [a-λ,a+λ]. (Period 2λ , Semiperiod λ)

$$\int_{a-\lambda}^{a+\lambda} \cos(uk.x).dx \overset{k=1,2,3,..}{=} \int_{a-\lambda}^{a+\lambda} \sin(uk.x).dx \overset{k=0,1,2,3...}{=} 0 \qquad uk=k.\frac{\pi}{\lambda}$$

$$\int_{a-\lambda}^{a+\lambda} \sin(uk.x).\cos(u\mu.x).dx \overset{k \# \mu}{=} \int_{a-\lambda}^{a+\lambda} \cos(uk.x).\cos(u\mu.x).dx \overset{k \# \mu}{=} 0$$

$$\int_{a-\lambda}^{a+\lambda} \cos^2(uk.x).dx = \int_{a-\lambda}^{a+\lambda} \frac{1+\cos(2uk.x)}{2}.dx \begin{array}{l} = 2\lambda \quad iff \ k=0 \\ = \lambda \quad iff \ k\#0 \end{array}$$

$$\int_{a-\lambda}^{a+\lambda} \sin^2(uk.x).dx = \int_{a-\lambda}^{a+\lambda} \frac{1-\cos(2uk.x)}{2}.dx \begin{array}{l} = 0 \quad iff \ k=0 \\ = \lambda \quad iff \ k\#0 \end{array}$$

Performing **linear transformation** every domain [a-λ,a+λ]
can be converted into a domain [-π,π]

A. From the fixed x in (a-λ,a+λ) to the fixed x' in (-π,π)
B. From the fixed x in (-λ,λ) to the fixed x' in (-π,π) **a=0**

Fixed x
a-λ< x <a+λ
|
x=a+ξ.λ
-1< ξ <1

Fixed x'
-π< x' <π
|
x'=ξ.π
ξ=(x-a)/λ

Thus: $x = a + \dfrac{\lambda}{\pi}.x'$

$$\int_{-\pi}^{\pi} \cos(k.x).dx \overset{k=1,2,3,..}{=} \int_{-\pi}^{\pi} \sin(k.x).dx = \int_{-\pi}^{\pi} \begin{bmatrix} \sin(k.x).\cos(\mu.x) \\ or \\ \cos(k.x).\cos(\mu.x) \end{bmatrix}.dx \overset{k \# \mu \quad \mu=1,2,3,.}{=} 0$$

$$\int_{-\pi}^{\pi} \cos^2(k.x).dx = \int_{-\pi}^{\pi} \frac{1+\cos(2k.x)}{2}.dx \begin{array}{l} = 2\pi \quad iff \ k=0 \\ = \pi \quad iff \ k\#0 \end{array}$$

$$\int_{-\pi}^{\pi} \sin^2(k.x).dx = \int_{-\pi}^{\pi} \frac{1-\cos(2k.x)}{2}.dx \begin{array}{l} = 0 \quad iff \ k=0 \\ = \pi \quad iff \ k\#0 \end{array}$$

$$Sf(x) = \frac{1}{\lambda} . \int_{-\lambda}^{\lambda} f(t) . \left[\frac{1}{2} + \sum_{k=1}^{+\infty} \cos[uk.(t-x)] \right].dt \qquad uk=k.\frac{\pi}{\lambda} \quad \begin{array}{c} x \ fixed \\ | \\ -\lambda<x<\lambda \end{array}$$

$$= \frac{1}{\pi} . \int_{-\pi}^{\pi} f(\frac{\lambda}{\pi}.t) . \left[\frac{1}{2} + \sum_{k=1}^{+\infty} \cos[k.(t-x')] \right].dt \qquad \begin{array}{c} x' fixed \\ | \\ -\pi<x'<\pi \end{array}$$

The classical approach to Fourier's series

$$f(x) = Sf(x) + rn(x) \quad\Big|\quad Sf(x) = [ao] + \sum_{k=1}^{n \to +\infty} ak.\cos(uk.x) + \sum_{k=1}^{n \to +\infty} bk.\sin(uk.x) \qquad uk = k.(\pi/\lambda)$$

Sf(x) = Trigonometric serial expansion of the function f(x)
rn(x) = Remainder function
Fourier's coefficients [ao], ak, bk k=1,2,3,...,n (n->+oo)

For the Fourier's coefficients the integrals of rn(x) vanish

$$\int_{-\lambda}^{\lambda} f(x).\cos(uk.x).dx - \int_{-\lambda}^{\lambda} Sf(x).\cos(uk.x).dx = \int_{-\lambda}^{\lambda} rn(x).\cos(uk.x).dx$$

--------=0----------

$$\int_{-\lambda}^{\lambda} f(x).\sin(uk.x).dx - \int_{-\lambda}^{\lambda} Sf(x).\sin(uk.x).dx = \int_{-\lambda}^{\lambda} rn(x).\sin(uk.x).dx$$

By orthogonality properties of trigonometric integrals

For k=0 $$\int_{-\lambda}^{\lambda} f(x).dx - \int_{-\lambda}^{\lambda} [ao].dx = \int_{-\lambda}^{\lambda} rn(x).dx = 0$$

k=1

$$\int_{-\lambda}^{\lambda} f(x).\cos(u1.x).dx - \int_{-\lambda}^{\lambda} a1.\cos^2(u1.x).dx = \int_{-\lambda}^{\lambda} rn(x).\cos(u1.x).dx$$

--------=0----------

$$\int_{-\lambda}^{\lambda} f(x).\sin(u1.x).dx - \int_{-\lambda}^{\lambda} b1.\sin^2(u1.x).dx = \int_{-\lambda}^{\lambda} rn(x).\sin(u1.x).dx$$

k=2

$$\int_{-\lambda}^{\lambda} f(x).\cos(u2.x).dx - \int_{-\lambda}^{\lambda} a2.\cos^2(u2.x).dx = \int_{-\lambda}^{\lambda} rn(x).\cos(u2.x).dx$$

--------=0----------

$$\int_{-\lambda}^{\lambda} f(x).\sin(u2.x).dx - \int_{-\lambda}^{\lambda} b2.\sin^2(u2.x).dx = \int_{-\lambda}^{\lambda} rn(x).\sin(u2.x).dx$$

etc for k=3,4,5,6,..,n (n->+oo)

Fourier's coefficients
For which the integrals of rn(x) vanish (become zero)

	By independence of notation
$[ao] = \dfrac{1}{2.\lambda} \cdot \displaystyle\int_{-\lambda}^{\lambda} f(x).dx$ $uk=k.(\pi/\lambda)$ $ak = \dfrac{1}{\lambda} \cdot \displaystyle\int_{-\lambda}^{\lambda} f(x).cos(uk.x).dx$ $bk = \dfrac{1}{\lambda} \cdot \displaystyle\int_{-\lambda}^{\lambda} f(x).sin(uk.x).dx$	$[ao] = \dfrac{1}{\lambda} \cdot \displaystyle\int_{-\lambda}^{\lambda} \dfrac{1}{2}.f(t).dt$ $ak = \dfrac{1}{\lambda} \cdot \displaystyle\int_{-\lambda}^{\lambda} f(t).cos(uk.t).dt$ $bk = \dfrac{1}{\lambda} \cdot \displaystyle\int_{-\lambda}^{\lambda} f(t).sin(uk.t).dt$

The convergence theorem
for the appropriate selected f(x)   ie rn(x)->0

$$n->+oo \quad uk=k.(\pi/\lambda)$$

$$Sf(x) = \frac{1}{\lambda}.\int_{-\lambda}^{\lambda} f(t).\left[\frac{1}{2} + \sum_{k=1}^{n} cos[uk.(t-x)]\right].dt = \frac{f(x-)+f(x+)}{2} = f(x) \quad *$$

* Iff f(x) continuous at the specific fixed x

$$cos[uk.(t-x)] = cos(uk.x).cos(uk.t) + sin(uk.x).sin(uk.t)$$

t is the variable in [-λ,λ],while x is fixed somewhere in (-λ,λ)

The appropriate selected f(x)

1. The function f(x) is continuous in (-λ,0) and in (0,λ)

2. At x=0 this f(x) is either continuous or piecewise continuous

3. The function f(x) is considered as periodically continued on the whole real axis x (from -oo to +oo) with period 2λ.

4. By periodicity: f(λ+)=f(-λ+) and f(-λ-)=f(λ-)

Or the f(x) is a smooth curve over the semiperiod (0,λ)
This f(x) is considered odd or even function in (-λ,λ).
And the 2. 3. and 4. of the above.

An odd function f(t) by an even function cos(uk.t) makes an odd function f(t).cos(uk.t),which drops out in [-λ,λ].

An even function f(t) by an odd function sin(uk.t) makes an odd function f(t).sin(uk.t),which drops out in [-λ,λ].

Fourier's series - Formulas for certain f(x)

1.a Rectangular curve:

f(x)=1	O< x <π

$$Sf(O) = \frac{f(O-) + f(O+)}{2} = O$$

$$Sf(x) = \frac{2}{\pi} . \int_{O}^{\pi} \left[\sum_{k=1}^{n \to +\infty} \sin(k.x) . \sin(k.t) \right] . dt$$

$$Sf(x) = \frac{4}{\pi} . \left[\sin x + \frac{\sin 3x}{3} + \frac{\sin 5x}{5} + .. \right]$$

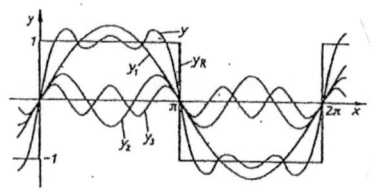

1.b Rectangular curve:

f(x)=a	O\< x <π/2	Sf(π/2)=O	f(x)=-a	π/2< x \<π

$$Sf(x) = \frac{2}{\pi} . \int_{O}^{\pi/2} a . \left[\frac{1}{2} + \sum_{k=1}^{n \to +\infty} \cos(k.x) . \cos(k.t) \right] . dt$$

$$Sf(x) = \frac{4.a}{\pi} . \left[\cos x - \frac{\cos 3x}{3} + \frac{\cos 5x}{5} - .. \right]$$

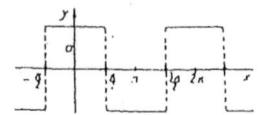

2.a Rectangular impulse of first kind:

f(x)=0	0< x <b		f(x)=a	b < x < π-b		f(x)=0	π-b< x <π

$$0< b <\pi$$

$$Sf(x) = \frac{2}{\pi} \int_{b}^{\pi-b} a.\left[\sum_{k=1}^{+\infty} \sin(k.x).\sin(k.t) \right].dt$$

$$Sf(x) = \frac{4.a}{\pi}.\left[\frac{\cos b}{1}.\sin x + \frac{\cos 3b}{3}.\sin 3x + \frac{\cos 5b}{5}.\sin 5x + .. \right]$$

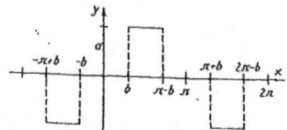

2.b Rectangular impulse of second kind:

f(x)=a	0 < x < c		f(x)=0	c< x <π

$$Sf(x) = \frac{2.a}{\pi}.\left[\frac{c}{2} + \frac{\sin c}{1}.\cos x + \frac{\sin 2c}{2}.\cos 2x + \frac{\sin 3c}{3}.\cos 3x + .. \right]$$

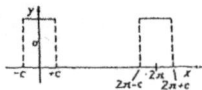

3. Sawtooth curve:

$$f(x)=\frac{2.a}{\pi}.x - a \quad | 0 < x <\pi$$

f(x)=(a/π).x - a	0< x \<π		f(0+)=-a	f(π)=0
f(x)=(a/π).x + a	-π\< x < 0		f(0-)= a	f(-π)=0

$$f(t) = (a/\pi).(t-\pi)$$

$$Sf(x) = \frac{2}{\pi}.\int_{0}^{\pi} f(t).\sum_{k=1}^{+\infty} \sin(k.x).\sin(k.t).dt$$

$$Sf(x) = -\frac{2.a}{\pi}.\left[\sin x + \frac{\sin 2x}{2} + \frac{\sin 3x}{3} +... \right]$$

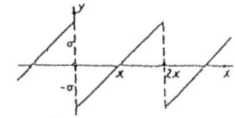

4. Triangular curve:

$$f(\chi) = \frac{2 \cdot a}{\pi} \cdot x \quad \boxed{0 < x \le \frac{\pi}{2}} \qquad f(x) = -\frac{2 \cdot a}{\pi} \cdot x + 2a \quad \boxed{\frac{\pi}{2} < x < \pi}$$

$$Sf(x) = \frac{8 \cdot a}{\pi^2} \cdot \left[\frac{\sin x}{1^2} - \frac{\sin 3x}{3^2} + \frac{\sin 5x}{5^2} - \cdots \right]$$

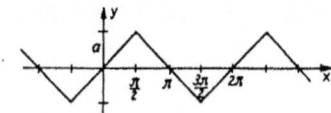

5. Triangular impulse:

$$f(x) = -\frac{a}{c} \cdot x + a \quad \boxed{0 < x < c} \qquad f(x) = 0 \quad \boxed{c < x < \pi}$$

$$Sf(x) = \frac{2}{\pi} \cdot \int_0^c f(t) \cdot \left[\frac{1}{2} + \sum_{k=1}^{+\infty} \cos(k \cdot x) \cdot \cos(k \cdot t) \right] \cdot dt$$

$$Sf(x) = \frac{a \cdot c}{2 \cdot \pi} + \frac{2 \cdot a}{\pi \cdot c} \cdot \left[\frac{1 - \cos c}{1^2} \cdot \cos x + \frac{1 - \cos 2c}{2^2} \cdot \cos 2x + \frac{1 - \cos 3c}{3^2} \cdot \cos 3x + \cdots \right]$$

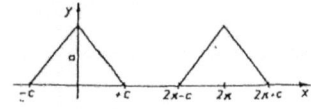

6.a Alternating current rectified in one direction

(Half waves of a cosine curve)

$f(x) = a.\cos x$	$0 < x < \pi/2$	$f(x) = 0$	$\pi/2 < x < \pi$

$$Sf(x) = \frac{2}{\pi}.\int_{0}^{\pi/2} a.\cos t.\left[\frac{1}{2} + \sum_{k=1}^{+\infty} \cos(k.x).\cos(k.t)\right].dt$$

$$\frac{a}{\pi}.\left[1 + \frac{\pi}{2}.\cos x + \frac{2}{1.3}.\cos 2x - \frac{2}{3.5}.\cos 4x + \frac{2}{5.7}.\cos 6x - \frac{2}{7.9}.\cos 8x + .\right]$$

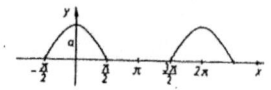

6.b Alternating current rectified in two directions

$f(x) = a.|\cos x|$	$0 < x < \pi$

$$Sf(x) = \frac{2}{\pi}.\int_{0}^{\pi} a.|\cos t|.\left[\frac{1}{2} + \sum_{k=1}^{+\infty} \cos(k.x).\cos(k.t)\right].dt$$

$$Sf(x) = \frac{2.a}{\pi}.\left[1 + \frac{2}{1.3}.\cos 2x - \frac{2}{3.5}.\cos 4x + \frac{2}{5.7}.\cos 6x - ..\right]$$

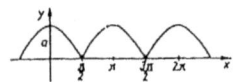

A unique approach to Fourier's series and integral

t is the variable , while x is fixed $(-\pi < x < \pi)$

$$\cos[k.(t-x)] = \cos(k.x).\cos(k.t) + \sin(k.x).\sin(k.t)$$

$$Sf(x) = \frac{1}{\pi} . \int_{-\pi}^{\pi} f(t).\left[\frac{1}{2} + \sum_{k=1}^{+\infty} \cos[k.(t-x)]\right].dt = \frac{f(x-)+f(x+)}{2} \;{}^* = f(x)$$

$\;^*$ Iff $f(x)$ is continuous at selected fixed x

$$Sf(x) = \frac{1}{\pi} . \int_{-\pi}^{\pi} f(t).\left[\int_{0}^{+\infty} \cos[k.(t-x)].dk\right].dt$$

Evaluated by the average value over the unit interval
Discrete variable k converted into a continuous one

$$Sf(x) = \frac{1}{\pi} . \int_{-\pi}^{\pi} f(t).\left[\frac{\sin[k.(t-x)]}{(t-x)}\Big|_{k=0}^{k+\infty}\right].dt \quad \boxed{\begin{array}{l}\text{By evaluation of}\\ \text{the integral of dk}\end{array}}$$

$$= \frac{1}{\pi}.\int_{-\pi}^{\pi} f(t).\frac{\sin[\omega.(t-x)]}{(t-x)}\overset{\omega->+\infty}{.}dt = \int_{-\pi}^{\pi} f(t).R(t-x).dt = \frac{f(x-)+f(x+)}{2}$$

$$Sf(x) = \frac{2}{\pi} . \int_{0}^{\pi} f(t).\left[\int_{0}^{+\infty}\left[\begin{array}{c}\cos(k.x).\cos(k.t)\\ \text{or}\\ \sin(k.x).\sin(k.t)\end{array}\right].dk\right].dt \quad \begin{array}{l}f(x)\ \textbf{even}\\ \\ f(x)\ \textbf{odd}\end{array}$$

$$= \frac{2}{\pi} . \int_{0}^{+\infty} . \int_{0}^{+\infty} f(t).\left[\begin{array}{c}\cos(k.x).\cos(k.t)\\ \text{or}\\ \sin(k.x).\sin(k.t)\end{array}\right].dk.dt \quad \begin{array}{l}f(x)\ \textbf{even}\\ \\ f(x)\ \textbf{odd}\end{array}$$

or dt.dk

Provided that: $\int_{-\infty}^{+\infty} |f(t)|.dt$ converges   The same interval of integration

The series of delta functions $P(t-x)$

$$P(t-x) = \frac{1}{\pi}.\left[\frac{1}{2} + \sum_{k=1}^{+\infty} \cos[k.(t-x)]\right] = \frac{1}{\pi}.\left[\frac{\sin[\omega.(t-x)]}{2.\sin[(t-x)/2]}\right] \quad \omega=->+\infty$$

The delta function $R(t-x) = \frac{1}{\pi}.\left[\dfrac{\sin[\omega.(t-x)]}{t-x}\right] \quad \omega->+\infty$
A continuous function

The graph of P(x) for n=10 and the graph of the function R(x)

As n->+oo the peaks tend to +oo and the small vibrations
become convergent to zero infinitesimals

$$P(x)=\frac{1}{\pi}.\left[\frac{1}{2}+\sum_{k=1}^{+\infty}\cos(k.x)\right]=\frac{1}{\pi}.\frac{\overset{\omega->+\infty}{\sin(\omega.x)}}{2.\sin(x/2)} \qquad R(x)=\frac{1}{\pi}.\frac{\overset{\omega->+\infty}{\sin(\omega.x)}}{x}$$

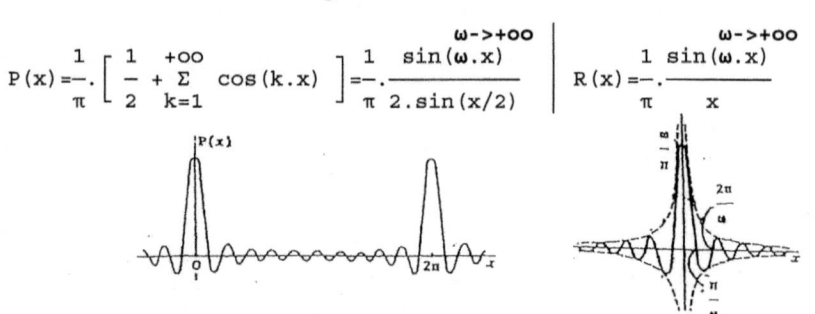

By translation of the axis we get the P(t-x) and R(t-x)

The heretic Dirac's isosceles triangle

. Extreme height -> +oo
. Very narrow base -> 0
. At the middle of this base is the fixed x
. The total aerea of this triangle is the unity $1 = \frac{1}{2} + \frac{1}{2}$
.Outside of this very narrow base ie for t#x the Dirac's delta
function , which is a continuous function , is an infinitesimal
so that the product of this delta function by the function f(t)
t#x is a **strong** infinitesimal , not contributing to the final
outcome of the integral.

For any k#0

$$\int_{-\pi}^{\pi}\cos(k.t).dt=\frac{\sin(k.t)}{k}\Big|_{k.t=-k.\pi}^{k.t=k.\pi}=0 \qquad \int_{-\pi}^{\pi}\sin(k.t).dt=\frac{-\cos(k.t)}{k}\Big|_{-k.\pi}^{k.\pi}=0$$

Orthogonality properties in the given domain

$$\int_{-\pi}^{\pi}\cos^2(k.t).dt = \frac{1}{k}.\left[\frac{1}{2}.[k.t + \sin(k.t).\cos(k.t)]\right]\Big|_{k.t=-k.\pi}^{k.t=k.\pi} = \pi$$

- -

$$\int_{-\pi}^{\pi}\sin^2(k.t).dt = \frac{1}{k}.\left[\frac{1}{2}.[k.t - \sin(k.t).\cos(k.t)]\right]\Big|_{k.t=-k.\pi}^{k.t=k.\pi} = \pi$$

The P(x) is the Fourier's expansion of the triangular impulse
as a->+oo & c->0 & a.c=1

Fourier's expansion of the triangular impulse,
an even function

$$f(x) = -\frac{a}{c}.x + a \quad \boxed{0<x<c} \qquad f(x)=0 \quad \boxed{c<x<\pi}$$

Principal domain $(-\pi,\pi)$
----
f(x) is even in $(-\pi,\pi)$

$$Sf(x)=\frac{2}{\pi}.\int_0^c f(t).\left[\frac{1}{2} + \sum_{k=1}^{+oo} \cos(k.x).\cos(k.t)\right].dt$$

The sine part
drops out
(even function)

$$Sf(x)=\frac{2}{\pi}.\left[-\frac{a\ c}{c\ 4}.2 + \frac{a.c}{2} - \frac{a}{c}.\left[\sum_{k=1}^{n} \frac{[\cos(k.c)-1]}{k^2}.\cos(k.x)\right]\right] \quad n->+oo$$

$$=\frac{a.c}{2.\pi} + \frac{2.a}{\pi.c}.\left[\sum_{k=1}^{n} \frac{1-\cos(k.c)}{k^2}.\cos(k.x)\right] \quad \boxed{n->+oo}$$

$$=\frac{a.c}{2.\pi} + \frac{2.a}{\pi.c}.\left[\frac{1-\cos c}{1^2}.\cos x + \frac{1-\cos 2c}{2^2}.\cos 2x + \frac{1-\cos 3c}{3^2}.\cos 3x +..\right]$$

as a->+oo & c->0 & a.c=1

By Taylor's series expansion:
of cos(k.c)

$$\lim_{c->0} \frac{1-\cos(k.c)}{k^2} = c^2/2$$

$$Sf(x) = \frac{1}{\pi}.\left[\frac{1}{2} + \sum_{k=1}^{n} \cos(k.x)\right] \overset{n->+oo}{} = \frac{1}{\pi}.\frac{\sin(\omega.x)}{2.\sin(x/2)} \overset{\omega->+oo}{} = P(x)$$

The graph of the triangular impulse

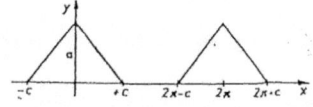

By translation of the axis (Setting t-x instead of x) **we get the series of delta functions** P(t-x).

Expansion over the semiperiod - Equivalent form

odd The cosine part drops out

$$Sf(x) = \frac{2}{\pi} . \int_0^\pi . \int_0^{+\infty} f(t).\sin(k.x).\sin(k.t).dk.dt = \frac{f(x-)+f(x+)}{2}$$

Transforming the discrete variable k into a continuous one
(By the average value over the unit interval)

The f(t) is the general formula of the smooth curve in (0,π)
t is the variable in [0,π],while x is fixed somewhere in (0,π)

odd f(t)

$$\frac{1}{2}.\left[\overset{k=0}{\underset{\text{---------=0----------}}{f(t).\sin(0.x).\sin(0.t)}} + \overset{k=1}{f(t).\sin(1.x).\sin(1.t)} \right]$$

$$+ \frac{1}{2}.\left[\overset{k=1}{f(t).\sin(1.x).\sin(1.t)} + \overset{k=2}{f(t).\sin(2.x).\sin(2.t)} \right]$$

$$+ \frac{1}{2}.\left[\overset{k=2}{f(t).\sin(2.x).\sin(2.t)} + \overset{k=3}{f(t).\sin(3.x).\sin(3.t)} \right]$$

$$+..$$

even The sine part drops out

$$Sf(x) = \frac{2}{\pi} . \int_0^\pi . \int_0^{+\infty} f(t).\cos(k.x).\cos(k.t).dk.dt = \frac{f(x-)+f(x+)}{2}$$

The f(t) is the general formula of the smooth curve in (0,π)
t is the variable in [0,π],while x is fixed somewhere in (0,π)

even f(t)

$$\frac{1}{2}.\left[\overset{k=0}{\underset{\text{-------=1--------}}{f(t).\cos(0.x).\cos(0.t)}} + \overset{k=1}{f(t).\cos(1.x).\cos(1.t)} \right]$$

$$+ \frac{1}{2}.\left[\overset{k=1}{f(t).\cos(1.x).\cos(1.t)} + \overset{k=2}{f(t).\cos(2.x).\cos(2.t)} \right]$$

$$+ \frac{1}{2}.\left[\overset{k=2}{f(t).\cos(2.x).\cos(2.t)} + \overset{k=3}{f(t).\cos(3.x).\cos(3.t)} \right]$$

$$+..$$

Fourier's integral - Modified presentation

From the appropriate function $f(x)$ to the series $Sf(x)$

such that: $\dfrac{f(x-)+f(x+)}{2} = Sf(x)$

Provided that $\displaystyle\int_{-\infty}^{+\infty} |f(t)|\,.dt$ **converges**

$$\int_{0}^{+\infty} \frac{\cos(a.x)}{b^2 + x^2}\,.dx = \frac{\pi}{2.b}.e^{-a.b} \qquad (a,b>0)$$

---

t is the variable , while x is fixed

$$e^{-s.x} = \frac{2}{\pi}.\int_{0}^{+\infty} e^{-s.t}.\cos(k.t).dt \;.\; \int_{0}^{+\infty} \cos(k.x).dk \qquad \begin{array}{l} s>0 \\ k>0 \end{array}$$

$$= \frac{2}{\pi}.\int_{0}^{+\infty} \left[\frac{s}{s^2 + k^2} \right].\cos(k.x).dk = \frac{2}{\pi}.\sum_{k=0}^{+\infty} \left[\frac{s}{s^2 + k^2} \right].\cos(k.x)$$

$$= \text{--------}=Sf(x)\text{-----------}$$

By the average value over the unit interval

Therefore:

$$\int_{0}^{+\infty} \left[\frac{1}{s^2 + k^2} \right].\cos(k.x).dk = \frac{\pi}{2.s}.e^{-s.x} \qquad \begin{array}{l} s>0 \\ k>0 \end{array}$$

The above is equivalent to the given integral

Fourier's integral - Modified presentation

From the appropriate function $f(x)$ **to the series** $Sf(x)$

such that: $\dfrac{f(x-)+f(x+)}{2} = Sf(x)$

Provided that $\displaystyle\int_{-\infty}^{+\infty} |f(t)|\,.dt$ **converges**

$$\int_{0}^{+\infty} \frac{t^3\,.\sin(t.x)}{t^4 + 4}\,.dt = \frac{\pi}{2}.e^{-x}\,.\cos x \qquad x>0$$

---

t is the variable , while x is fixed

$$e^{-x}.\cos x = \frac{2}{\pi}.\int_{0}^{+\infty} e^{-t}\,.\cos t.\sin(k.t)\,.dt\ .\int_{0}^{+\infty}\sin(k.x)\,.dk \qquad \begin{array}{l} x>0 \\ k>0 \end{array}$$

$$\begin{array}{l} 0< x <\pi \\ x \text{ is fixed} \end{array}\Bigg| \quad =\frac{2}{\pi}.\int_{0}^{+\infty} e^{-t}\ .\left[\frac{\sin[(k+1).t]+\sin[(k-1).t]}{2}\right].dt.\int_{0}^{+\infty}\sin(k.x)\,.dk$$

$$\Bigg| \quad = \frac{2}{\pi}.\int_{0}^{+\infty} \frac{1}{2}.\left[\frac{k+1}{1 + (k+1)^2} + \frac{k-1}{1 + (k-1)^2}\right].\sin(k.x)\,.dk$$

$$\Bigg| \quad = \frac{2}{\pi}.\int_{0}^{+\infty} \left[\frac{k^3}{k^4 + 4}\right].\sin(k.x)\,.dk = \frac{2}{\pi}.\sum_{k=1}^{+\infty}\left[\frac{k^3}{k^4 +4}\right].\sin(k.x)$$

$$= Sf(x)$$

By the average value over the unit interval starting from k=0

Therefore:

$$\int_{0}^{+\infty} \left[\frac{k^3}{k^4 + 4}\right].\sin(k.x)\,.dk = \frac{\pi}{2}.e^{-x}\,.\cos x \qquad x>0$$

The above is equivalent to the given integral

Useful formulas derived from Fourier' series

$$\frac{(x-)+(x+)}{2} = x = -2 . \sum_{k=1}^{+\infty} \frac{(-1)^k}{k} . \sin(k.x)$$

$$\frac{\pi}{2} = 2 . \left[1 - \frac{1}{3} + \frac{1}{5} - \frac{1}{7} + .. \right.$$

$$x = \frac{2}{\pi} . \left[\frac{\pi^2}{4} - 2 . \sum_{k=1}^{n} \frac{1}{k^2} \right] . \cos(k.x) \quad k \text{ odd}$$

$$x = \frac{\pi}{2} - \frac{4}{\pi} . \left[\frac{\cos x}{1^2} + \frac{\cos 3x}{3^2} + \frac{\cos 5x}{5^2} + .. \right]$$

Euler's formula setting x=0

$$\frac{\pi^2}{8} = 1 + \frac{1}{3^2} + \frac{1}{5^2} + \frac{1}{7^2} + ..$$

$$x = \frac{\pi^2}{3} + 4 . \sum_{k=1}^{+\infty} \frac{(-1)^k}{k^2} . \cos(k.x) = \frac{\pi^2}{3} + 4 . \sum_{k=1}^{+\infty} \frac{\cos(k.\pi)}{k^2} . \cos(k.x)$$

Setting x=0 and taking into consideration the Euler's formula

k odd	k even	k odd and even
$\sum_{k=1}^{+\infty} (1/k^2) = \pi^2/8$	$\sum_{k=2}^{+\infty} (1/k^2) = \pi^2/24$	$\sum_{k=1}^{+\infty} (1/k^2) = \pi^2/6$

Sums of some Fourier's Series - Slow convergent series

1.
$$\sum_{k=1}^{n \to +\infty} \frac{\sin(k.x)}{k} = \sigma_0(x) = \frac{\pi-x}{2}$$

Principal domain: $0 < x < 2.\pi$

By integration of $\sigma_0(x)$

2.
$$-\sum_{k=1}^{n} \frac{\cos(k.x)}{k^2} = -\sigma_1(x) = \frac{6.\pi.x - 3.x^2 - 2.\pi^2}{12}$$

$0 \backslash < x \backslash < 2.\pi$

By integration of $\sigma_1(x)$

3.
$$\sum_{k=1}^{n} \frac{\sin(k.x)}{k^3} = \sigma_2(x) = \frac{2.\pi^2.x - 3.\pi.x^2 + x^3}{12}$$

$0 \backslash < x \backslash < 2.\pi$

4.
$$\sum_{k=1}^{n} \frac{\cos(k.x)}{k} = -\ln(2.\sin\frac{x}{2})$$

$0 < x < 2.\pi$

$$\sum_{k=1}^{n} \frac{\cos(k.\pi)}{k} = -\ln 2 = -0.69314718..$$

By integration of 4.

5.
$$\sum_{k=1}^{n} \frac{\sin(k.x)}{k^2} = -\int_{0}^{x} \ln(2.\sin\frac{x}{2}).dx$$

$0 \backslash < x \backslash < 2.\pi$

By integration of 5.

6.
$$\sum_{k=1}^{n} \frac{\cos(k.x)}{k^3} = \int_{0}^{x} dx . \int_{0}^{x} \ln(2.\sin\frac{x}{2}).dx + \sum_{k=1}^{n} (1/k^3)$$

$0 \backslash < x \backslash < 2.\pi$

$$= 1.202056903..$$

Various proofs related to Fourier's series and integral

$$\text{The proof that } \int_0^{+\infty} \frac{\sin x}{x}\,.dx = \frac{\pi}{2}$$

The integrand function is even

$$= \int_0^{+\infty} e^{-t}\,.dt \cdot \int_0^{+\infty} \frac{\sin x}{x}\,.dx \qquad\qquad \int_0^{+\infty} e^{-t}\,.dt = 1$$

$$= \int_0^{+\infty} e^{-t}\,.dt \cdot \int_0^{+\infty} \frac{\sin(t.x)}{t.x}\,.d(t.x) \qquad\qquad \begin{array}{c}\text{By independence}\\ \text{of notation}\end{array}$$

$$= \int_0^{+\infty} e^{-t}\,.\sin(x.t)\,.dt \cdot \int_0^{+\infty} \frac{1}{x}\,-.dx \qquad \int_0^{+\infty} e^{-s.t}\,.\sin(u.t)\,.dt = \frac{u}{s^2 + u^2}$$

$$= \int_0^{+\infty} \left[\frac{x}{1 + x^2}\right] \frac{1}{x}\,.dx = \arctan x \Big|_{x=0}^{x=+\infty} = \frac{\pi}{2} \qquad s>0$$

- -

Therefore: $\frac{1}{\pi}.\int_{-\infty}^{+\infty} \frac{\sin x}{x}\,.dx = 1$ **The integrand function is even**

By independence of notation

$$\frac{1}{\pi}.\int_{-\infty}^{+\infty} \frac{\sin(k.x)}{k.x}\,.d(k.x) = \frac{1}{\pi}.\int_{-\infty}^{+\infty} k.\frac{\sin(k.x)}{k.x}\,.dx = \frac{1}{\pi}.\int_{-\infty}^{+\infty} \frac{\sin(k.x)}{x}\,.dx = 1 = \frac{1}{2}+\frac{1}{2}$$

---

k<1 (k>0)
Shrinking in y direction and stretching in x direction by the
same factor does not change the value of the improper integral
k>1
Stretching in y direction and shrinking in x direction by the
same factor does not change the value of the improper integral
**For k->+oo the integrand function is converted
into the delta function R(x)**

By translation of the axis
Setting (t-x) instead of x , we get the delta function R(t-x)

$$\textbf{The delta function } R(t-x) = \frac{1}{\pi}.\left[\frac{\sin[\omega.(t-x)]}{t-x}\right] \quad \omega->+\infty$$

t is the variable , while x is fixed (-π<x<π)

$$\text{Another proof that } \int_0^{+\infty} \frac{\sin x}{x}.dx = \frac{\pi}{2}$$

Improper Integration
based on differentiation under the integral sign

$$s>0$$

$$I(s,b) = \int_0^{+\infty} \frac{e^{-s.t}.\sin(b.t)}{t}.dt = \frac{\pi}{2} - \arctan\frac{s}{b} = \arctan\frac{b}{s}$$

$$(50.)$$

$$= (\pi/2 - v) \quad = \quad u$$

For s->0 the proof is obvious

$$\sqrt{s^2 + b^2}$$

$$v = \frac{\pi}{2} - u$$

$$\int \frac{ds}{s^2 + b^2} = \frac{1}{b}.\arctan\frac{s}{b} + C$$

$$\frac{dI(s,b)}{ds} = \int_0^{+\infty} e^{-s.t}.(-t).\frac{\sin(b.t)}{t}.dt = -\int_0^{+\infty} e^{-s.t}.\sin(b.t).dt$$

$$= -\frac{b}{(s^2 + b^2)}$$

- -

$$I(s,b) = -\int b.\left[s^2 + b^2\right]^{-1}.ds = -b.\left[\frac{1}{b}.\arctan\frac{s}{b}\right] + C$$

$$s->+\infty$$

$$I(s,b) = \int_0^{+\infty} \frac{e^{-s.t}.\sin(b.t)}{t}.dt = 0 = -\arctan\frac{s}{b} + C = -\frac{\pi}{2} + C$$

$$s->+\infty$$

$$\textbf{Therefore: } C = \frac{\pi}{2}$$

- Reference 8. -

Dirac's principle

$$2. \int_0^{+\infty} e^{-x^2}.dx = \int_{-\infty}^{+\infty} e^{-x^2}.dx = \sqrt{\pi} \;\Rightarrow\; \int_{-\infty}^{+\infty} \sqrt{s}.e^{-s.t^2}.dt = \sqrt{\pi} \quad (s>0)$$

Even integrand function $k.f(k.x)$

---

k<1 (k>0)
Shrinking in y direction and stretching in x direction by the same factor does not change the value of the improper integral

k>1
Stretching in y direction and shrinking in x direction by the same factor does not change the value of the improper integral

**For k->+oo all those functions can be converted
into the delta function δ(x) or δ(t-x)**

Examples of even functions bounded for all x ie -oo<x<+oo
which can be converted into delta functions

$$\varphi 1(x) = \frac{1}{2.\left[1+x^2\right]^{3/2}}$$

$$\varphi 2(x) = \frac{1}{\pi}.\frac{1}{1+x^2}$$

$$\varphi 3(x) = \frac{1}{\sqrt{\pi}}.e^{-x^2}$$

$$\varphi 4(x) = \frac{1}{\pi}.\frac{\sin x}{x}$$

The common property of those funtions: $\int_{-\infty}^{+\infty} \varphi i(x).dx = 1$ i=1,2,3,4

Notice also: $\int_{-\infty}^{+\infty} k.\varphi i(k.x).dx = 1$ (i=1,2,3,4) $\boxed{k>0}$

$$\int_{-\infty}^{+\infty} f(x).\delta(x).dx = \frac{f(0-)+f(0+)}{2} \quad \& \quad \int_{-\infty}^{+\infty} f(t).\delta(t-x).dt = \frac{f(x-)+f(x+)}{2}$$

Provided that: $\int_{-\infty}^{+\infty} |f(t)|.dt$ converges

Absolute convergence implies convergence

The convergence of the integral   **is quarranteed iff**

$$\int_{-\infty}^{+\infty} f(t).\sin(u.t).dt = M(u) \qquad \int_{-\infty}^{+\infty} |f(t)|.dt \text{ converges}$$

Graphs of the following even functions bounded for all x
$$-\infty < x < +\infty$$

$$\varphi 1(x) = \frac{1}{2.\left[1+x^2\right]^{3/2}}$$

$$\varphi 2(x) = \frac{1}{\pi}.\frac{1}{1+x^2}$$

$$\varphi 3(x) = \frac{1}{\sqrt{\pi}}.e^{-x^2}$$

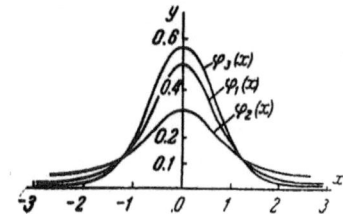

The $\varphi 1(x)$ is the second derivative of $\sqrt{x^2 + a^2}$ for a=1

Bessel's inequality

By this inequality the convergence theorem of Fourier's series
for the appropriate selected $f(x)$ is prooven.
- The Dirac's approach to Fourier's series is by far superior -

$$\frac{1}{\pi} . \int_{-\pi}^{\pi} [f(x)]^2 . dx > \frac{1}{\pi} . \int_{-\pi}^{\pi} [Sf(x)]^2 . dx = \frac{a_0^2}{2} + \sum_{k=1}^{n \to \infty} [a_k^2 + b_k^2]$$

By the comparisson to the convergent integral the above series
converges, meaning that **as n->+oo** the a_n term and b_n term tend
to zero and that the limit of the ratio of $|a_{n+1}|/|a_n|$ terms
as n->+oo is less than one. (The ratio test)

The convergence however is slow

There are techniques of accelerating this convergence
and thus saving a lot of computations

The proof of Bessel's inequality

From $f(x) = Sg(x) + r_n(x)$ we get: $\int_{-\pi}^{\pi} [f(x)]^2 . dx = \int_{-\pi}^{\pi} [Sf(x) + r_n(x)]^2 . dx$

$$\int_{-\pi}^{\pi} [f(x)]^2 . dx = \int_{-\pi}^{\pi} \left[[Sf(x)]^2 + [r_n(x)]^2 + 2.Sf(x).r_n(x) \right] . dx$$

For the Fourier's coefficients the integrals of $r_n(x)$ vanish.

$$= \int_{-\pi}^{\pi} \left[[Sf(x)]^2 + [r_n(x)]^2 \right] . dx$$

By orthogonality properties of trigonometric integrals

$$= \int_{-\pi}^{\pi} \left[[a_0]^2 + \sum_{k=1}^{+\infty} a_k^2 . \cos^2 (k.x) + \sum_{k=1}^{+\infty} b_k^2 . \sin^2 (k.x) + [r_n(x)]^2 \right] . dx$$

$$\frac{1}{\pi} . \int_{-\pi}^{\pi} \sum_{k=1}^{+\infty} a_k^2 . \cos^2 (k.x) . dx = \frac{1}{\pi} . \int_{-\pi}^{\pi} \sum_{k=1}^{+\infty} a_k^2 . \frac{1 + \cos(2k.x)}{2} . dx = \sum_{k=1}^{+\infty} a_k^2$$

$$\frac{1}{\pi} . \int_{-\pi}^{\pi} [a_0]^2 . dx = 2.[a_0]^2 = a_0^2 / 2 \qquad \boxed{a_0 = [a_0]^2 / 2} \qquad = a_0^2 / 2$$

$$\frac{1}{\pi} . \int_{-\pi}^{\pi} \sum_{k=1}^{+\infty} b_k^2 . \sin^2 (k.x) . dx = \frac{1}{\pi} . \int_{-\pi}^{\pi} \sum_{k=1}^{+\infty} b_k^2 . \frac{1 - \cos(2k.x)}{2} . dx = \sum_{k=1}^{+\infty} b_k^2$$

Harmonic Analysers

Computers approximatelly converting a graph not
given in analytical form into Fourier's series

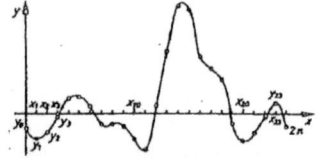

The domain of the graph is divided by the computer in 2.m equal
parts.   The m must be a multiple of 4.   This fascilitates the
calculations.Depending on the power of the computer the highest
possible m is selected.

The theoretical basis in its simplest form for m=12

A linear system of 24 equations with 24 unknowns
[ao],ak k=1,2,3,....,11
bk k=1,2,3,....,11,12

$$[ao] + \sum_{k=1}^{11} ak.\cos(uk.x1) + \sum_{k=1}^{11} bk.\sin(uk.x1) + b12.\sin(u12.x1) = y1$$

$$[ao] + \sum_{k=1}^{11} ak.\cos(uk.x2) + \sum_{k=1}^{11} bk.\sin(uk.x2) + b12.\sin(u12.x2) = y2$$

$$[ao] + \sum_{k=1}^{11} ak.\cos(uk.x3) + \sum_{k=1}^{11} bk.\sin(uk.x3) + b12.\sin(u12.x3) = y3$$

....

$$[ao] + \sum_{k=1}^{11} ak.\cos(uk.x23) + \sum_{k=1}^{11} bk.\sin(uk.x23) + b12.\sin(u12.x23) = y23$$

$$[ao] + \sum_{k=1}^{11} ak.\cos(uk.x24) + \sum_{k=1}^{11} bk.\sin(uk.x24) + b12.\sin(u12.x24) = y24$$

Sf(x)=Trigonometric serial expansion of the given graph

$$Sf(x) = [ao] + \sum_{k=1}^{11} ak.\cos(uk.x) + \sum_{k=1}^{12} bk.\sin(uk.x) \quad \boxed{0< x <2\lambda}$$

In experienced hands those calculations require several hours
By the appropriate computers fractions of the second

Sums of some Fourier's Series - Slow convergent series
- The proofs -

1. $\displaystyle\sum_{k=1}^{n} \frac{\sin(k.x)}{k} = \sigma_0(x) = \frac{\pi-x}{2}$ $n\to+\infty$	**Principal domain** $0< x <2.\pi$

By integration of $\sigma_0(x)$

2. $\displaystyle -\sum_{k=1}^{n} \frac{\cos(k.x)}{k^2} = -\sigma_1(x) = \frac{6.\pi.x - 3.x^2 - 2.\pi^2}{12}$	$0\backslash<x\backslash<2.\pi$

By integration of $\sigma_1(x)$

3. $\displaystyle \sum_{k=1}^{n} \frac{\sin(k.x)}{k^3} = \sigma_2(x) = \frac{2.\pi^2.x - 3.\pi.x^2 + x^3}{12}$	$0\backslash<x\backslash<2.\pi$

4. $\displaystyle \sum_{k=1}^{n} \frac{\cos(k.x)}{k} = -\ln(2.\sin\frac{x}{2})$	$0< x <2.\pi$
$\displaystyle \sum_{k=1}^{n} \frac{\cos(k.\pi)}{k} = -\ln 2 = -0.69314718..$	

By integration of 4.

5. $\displaystyle \sum_{k=1}^{n} \frac{\sin(k.x)}{k^2} = -\int_0^x \ln(2.\sin\frac{x}{2}).dx$	$0\backslash<x\backslash<2.\pi$

By integration of 5.

6. $\displaystyle \sum_{k=1}^{n} \frac{\cos(k.x)}{k^3} = \int_0^x dx . \int_0^x \ln(2.\sin\frac{x}{2}).dx + \underbrace{\sum_{k=1}^{n} (1/k^3)}_{=1.202056903..}$	$0\backslash<x\backslash<2.\pi$

1.
$$\sum_{k=1}^{n} \frac{\sin kx}{k} = \frac{\sin x}{1} + \frac{\sin 2x}{2} + \frac{\sin 3x}{3} + .. = oo(x) = \frac{\pi - x}{2}$$

Principal domain

0< x <2.π

The graph of oo(x)

$y=\frac{\pi - x}{2}$

$$oo(t) = \frac{\pi - t}{2}$$

$$Sf(x) = \frac{1}{\pi}. \int_{0}^{2.\pi} f(t). \left[\sum_{k=1}^{n} \sin(k.x).\sin(k.t) \right].dt = \frac{f(x-)+f(x+)}{2}$$

(n->+oo above the sum)

$$\int_{0}^{2.\pi} \frac{t}{2}.\sin(k.t).dt = -\frac{\pi}{k}$$

- -

$$= -\frac{1}{k} \int \frac{t}{2}.d\cos(k.t) = -\frac{1}{k}. \left[\frac{t}{2}.\cos(k.t) \Big|_{0}^{2.\pi} - \int_{0}^{2.\pi} \cos(k.t).d\frac{t}{2} \right]$$

$$= -\frac{1}{k}. \left[\frac{t}{2}.\cos(k.t) \Big|_{0}^{2.\pi} - \frac{1}{2}.\frac{\sin(k.t)}{k} \Big|_{0}^{2.\pi} \right]$$

$$-\frac{x}{2} = \frac{1}{\pi}. \int_{0}^{2.\pi} -\frac{t}{2}. \left[\sum_{k=1}^{n} \sin(k.x).\sin(k.t) \right].dt = \sum_{k=1}^{n} \frac{\sin(k.x)}{k}$$

$$\int_{0}^{2.\pi} \frac{\pi}{2}.\sin(k.t).dt = \frac{\pi}{2}.(-\frac{1}{k}). \int_{t=0}^{t=2.\pi} d\cos(k.t) = 0$$

2. By integration of $\sigma_0(x)$	Principal domain

$$\sum_{k=1}^{n} \frac{\cos kx}{k^2} = \frac{\cos x}{1} + \frac{\cos 2x}{4} + \frac{\cos 3x}{9} + .. = -\sigma_1(x) = \frac{6\pi x - 3x^2 - 2\pi^2}{12}$$

$0\backslash <x\backslash <2.\pi$

1.	$n \to +\infty$	Principal domain

$$\sum_{k=1}^{n} \frac{\sin kx}{k} = \frac{\sin x}{1} + \frac{\sin 2x}{2} + \frac{\sin 3x}{3} + .. = \sigma_0(x) = \frac{\pi - x}{2}$$

$0< x <2.\pi$

$$\int_{0}^{x} \sum_{k=1}^{n} \frac{\sin kt}{k} . dt = \int_{t=0}^{t=x} d\left[- \sum_{k=1}^{n} \frac{\cos kt}{k^2} \right] = - \sum_{k=1}^{n} \frac{\cos kx}{k^2} + \sum_{k-1}^{n} \frac{1}{k^2}$$

$$= - \sigma_1(x) + \pi^2/6$$

$$\int_{0}^{x} \underbrace{\frac{\pi - t}{2}}_{\sigma_0(t)} . dt = \left[\frac{\pi}{2}.t - t^2/4 \right]_{t=0}^{t=x} = \frac{\pi}{2}.x - x^2/4$$

Therefore:- $\sigma_1(x) = \dfrac{6\pi x - 3x^2 - 2\pi^2}{12}$

The graph of $\sigma_1(x)$

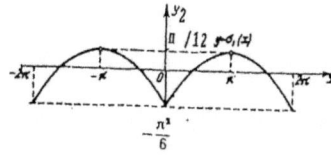

3. **By integration of σ1(x)**

$$\sum_{k=1}^{n} \frac{\sin kx}{k^3} = \frac{\sin x}{1} + \frac{\sin 2x}{8} + \frac{\sin 3x}{27} + \ldots = \sigma 2(x) = \frac{2\pi^2 x - 3\pi x^2 + x^3}{12}$$

Principal domain

$$0 \backslash < x \backslash < 2.\pi$$

$$\underbrace{\int_{0}^{x} \sum_{k=1}^{n} \frac{\cos kt}{k^2}.dt}_{\sigma 1(t)} = \int_{t=0}^{t=x} d\left[\sum_{k=1}^{n} \frac{\sin kt}{k^3}\right] = \underbrace{\sum_{k=1}^{n} \frac{\sin kx}{k^3}}_{\sigma 2(x)}$$

$$\underbrace{\int_{0}^{x} \frac{-6\pi t + 3t^2 + 2\pi^2}{12}.dt}_{\sigma 1(t)} = \left.\frac{-6\pi(t^2/2) + 3(t^3/3) + 2\pi^2 t}{12}\right|_{t=0}^{t=x}$$

$$\underbrace{\sigma 2(x)}_{} = \frac{2\pi^2 x - 3\pi x^2 + x^3}{12}$$

The proof that:

For every x belonging to (0,2π)

1.

$$P(x) = \frac{1}{\pi} \cdot \left[\frac{1}{2} + \overset{n\to+\infty}{\underset{k=1}{\overset{n}{\Sigma}} \cos(k.x)} \right] = \frac{1}{\pi} \cdot \frac{\overset{\omega\to+\infty}{\sin(\omega.x)}}{2.\sin(x/2)} \qquad \omega = n + \frac{1}{2} \qquad \boxed{n\to+\infty}$$

$$\overset{+\infty}{\underset{k=1}{\Sigma}} \frac{\cos(k.x)}{k} = -\ln\left(2.\sin\frac{x}{2}\right) \qquad \left| \; 0< x <2\pi \right.$$

Notice:

$$\overset{+\infty}{\underset{k=1}{\Sigma}} \frac{\cos(k.\pi)}{k} = -\ln2 = -0.69314718.. = -\left[1 - \frac{1}{2} + \frac{1}{3} - \frac{1}{4} + \frac{1}{5} - .. \right]$$

2.

$$\overset{+\infty}{\underset{k=1}{\Sigma}} \sin(k.x) = \frac{\cos(x/2) - \cos(\omega.x)}{2.\sin(x/2)} \qquad \omega = \left[n + \frac{1}{2} \right] \quad n\to+\infty \qquad 0< x <2\pi$$

The proof that for every x belonging to (0,2π)

$$P(x) = \frac{1}{\pi} \cdot \left[\frac{1}{2} + \sum_{k=1}^{n} \cos(k.x) \right]_{n \to +\infty} = \frac{1}{\pi} \cdot \frac{\sin(\omega.x)}{2.\sin(x/2)}_{\omega \to +\infty} \qquad \omega = n + \frac{1}{2} \qquad \boxed{n \to +\infty}$$

$$\left[\sin(0 + \frac{1}{2}).x \right] = \cos(0.x).\sin\frac{x}{2}$$

$$\left[\sin(1 + \frac{1}{2}).x - \sin(1 - \frac{1}{2}).x \right] = 2.\cos(1.x).\sin\frac{x}{2}$$

$$\left[\sin(2 + \frac{1}{2}).x - \sin(2 - \frac{1}{2}).x \right] = 2.\cos(2.x).\sin\frac{x}{2}$$

$$\left[\sin(3 + \frac{1}{2}).x - \sin(3 - \frac{1}{2}).x \right] = 2.\cos(3.x).\sin\frac{x}{2} \quad \text{etc}$$

$$_{n \to +\infty} \quad \sin(n + \frac{1}{2}).x = 2.\sin(\frac{x}{2}).\left[\frac{1}{2} + \sum_{k=1}^{+\infty} \cos(k.x) \right]$$

By LHospital's rule

$$\lim_{x \to 0} \frac{\sin(\omega.x)}{2.\sin(x/2)} = \lim_{x \to 0} \left[\frac{1}{2} + \sum_{k=1}^{+\infty} \cos(k.x) \right] = +\infty \qquad _{\omega \to +\infty}$$

And

$$\frac{1}{\pi} \cdot \int_{-\pi}^{\pi} \left[\frac{1}{2} + \sum_{k=1}^{+\infty} \cos(k.x) \right].dx = \int_{-\pi}^{\pi} P(x).dx = 1 = \frac{1}{2} + \frac{1}{2}$$

$$\frac{1}{\pi} \cdot \int_{-\pi}^{\pi} \left[\frac{1}{2} \right].dx = 1$$

$$\int_{-\pi}^{\pi} \sum_{k=1}^{+\infty} \cos(k.x).dx = 0 \qquad \int_{-\pi}^{\pi} \cos(k.x).dx = \frac{1}{k}.\sin(k.x) \Big|_{x=-\pi}^{x=\pi} = 0$$

The proof that for every x belonging in the domain (0,2π)

$$\sum_{k=1}^{+\infty} \sin(k.x) = \frac{\cos(x/2) - \cos(\omega.x)}{2.\sin(x/2)}$$

$$\omega = \left[n + \frac{1}{2} \right] \quad n \to +\infty \qquad 0 < x < 2\pi$$

From
$$\sum_{k=0}^{n} \cos\left[(k + \frac{1}{2}).x\right] = \sum_{k=1}^{n+1} \cos\left[(k - \frac{1}{2}).x\right]$$
we get:

$$\sum_{k=0}^{n} \cos\left[(k + \frac{1}{2}).x\right] = \sum_{k=0}^{n} \cos(k.x).\cos\frac{x}{2} - \sum_{k=0}^{n} \sin(k.x).\sin\frac{x}{2}$$

$$\sum_{k=1}^{n+1} \cos\left[(k - \frac{1}{2}).x\right] = \sum_{k=1}^{n+1} \cos(k.x).\cos\frac{x}{2} + \sum_{k=1}^{n+1} \sin(k.x).\sin\frac{x}{2}$$

$$\sum_{k=1}^{n+1} \cos\left[(k - \frac{1}{2}).x\right] = \begin{aligned} &\sum_{k=0}^{n} \cos(k.x).\cos\frac{x}{2} + \sum_{k=0}^{n} \sin(k.x).\sin\frac{x}{2} \\ &+ \cos[(n+1).x].\cos\frac{x}{2} + \sin[(n+1).x].\sin\frac{x}{2} \\ &- \cos(x/2) \end{aligned}$$

By substraction we get :

$$0 = -2.\sum_{k=0}^{n} \sin(k.x).\sin\frac{x}{2} + \cos\frac{x}{2} - \cos[(n+1).x].\cos\frac{x}{2} - \sin[(n+1).x].\sin\frac{x}{2}$$

$$0 = -2.\sum_{k=1}^{n} \sin(k.x).\sin\frac{x}{2} + \cos\frac{x}{2} - \cos\left[(n+1 - \frac{1}{2}).x\right]$$

The proof that:

$$\sum_{k=1}^{+\infty} \frac{\cos(k.x)}{k} = -\ln(2.\sin\frac{x}{2}) \qquad \Big| \quad 0< x <2\pi$$

$$\sum_{k=1}^{+\infty} \frac{\cos(k.\pi)}{k} = -\ln2 = -0.69314718.. = -\left[1 - \frac{1}{2} + \frac{1}{3} - \frac{1}{4} + \frac{1}{5} -..\right]$$

$$\int_x^\pi \sum_{k=1}^{+\infty} \sin(k.t).dt = -\sum_{k=1}^{+\infty} \frac{\cos(k.t)}{k}\Big|_{t=x}^{t=\pi} = -\sum_{k=1}^{+\infty} \frac{\cos(k.\pi)}{k} + \sum_{k=1}^{+\infty} \frac{\cos(k.x)}{k}$$

$$= \int_x^\pi \frac{\cos(t/2) - \cos(\omega.t)}{2.\sin(t/2)}.dt \qquad \boxed{\omega\text{->}+\infty}$$

$$= \ln(\sin\frac{t}{2})\Big|_{t=x}^{t=\pi} - \int_x^\pi \frac{\cos(\omega.t)}{2.\sin(t/2)}.dt = -\ln(\sin\frac{x}{2})$$

$$\int_x^\pi \frac{\cos(\omega.t)}{2.\sin(t/2)}.dt = \frac{1}{\omega}.\int_{t=x}^{t=\pi} [2.\sin(t/2)]^{-1}.d\sin(\omega.t) = \qquad \omega\text{->}+\infty$$

Integration by parts (IBP)

$$\boxed{\text{IBP}} = \frac{1}{\omega}.\left[\frac{\sin(\omega.t)}{2.\sin(t/2)}\Big|_{t=x}^{t=\pi} - \int_{t=x}^{t=\pi} \sin(\omega.t).d[2.\sin(t/2)]^{-1}\right]$$

$$= \frac{1}{\omega}.\left[\frac{\sin(\omega.t)}{2.\sin(t/2)}\Big|_{t=x}^{t=\pi} + \int_x^\pi \frac{\sin(\omega.t)}{2.\sin(t/2)}.\frac{\cos(t/2)}{2.\sin(t/2)}.dt\right] = 0$$

$$d[2.\sin(t/2)]^{-1} = -1.[2.\sin(t/2)]^{-2}.2.(1/2).\cos(t/2).dt$$

Notice that: $\quad \dfrac{1}{\pi}.\dfrac{\sin(\omega.x)}{2.\sin(x/2)} = P(x) = \dfrac{1}{\pi}.\left[\dfrac{1}{2} + \sum_{k=1}^{n} \cos(k.x)\right]$

$\omega\text{->}+\infty \qquad n\text{->}+\infty$

The function $P(x)/\omega$ in the domain $(0,2\pi)$ behaves as a very strong infinitesimal.

References:

1. **Higher Mathematics for beginners:**
    ----
    by Ya.B.Zeldovich
    (Mir Publishers Moscow 1973)

2. **Calculus with analytic geometry:**
    ----
    by Harley Flanders and Justin J Price
    (Academic Press 1978)

3. **A brief course of higher mathematics:**
    ----
    by V.A.Kudryavtsev and B.P.Demidovich
    (Mir Publisher's Moscow 1980)

4. **Concice Encyclopedia of Mathematics:**
    ----
    by W.Gellert,H.Kustner,M Hellwich,H Kastner
    (Van Nostrand Reinhold Company New York and other cities 1977)

5. **Computational Mathematics:**
    ----
    by B.P Demitovich and I.A.Maron
    (Mir publishers Moscow 1976)

6. **Advanced calculus:**
    ----
    by Leopold Flatto
    (The Wiiliams and Wilkins Company - Baltimore 1982)

7. **Mathematics Handbook for Science and Engineering:**
    ----
    by: Royal Lennart Rade and Bertil Westegren
    Fifth edition - **2004**
    Springer Verlag Publications Inc
    Berlin - Heidelberg - New York

8. **Mathematical methods for physicists and engineers:**
    ----
    by: Royal Eugene Collins - 2nd corrected edition
    Dover Publications Inc - Mineola New York - USA 1991

9. **Differential Equations:**

    A systems approach - by: Jack Goldberg - and Merle C.Potter
    Prentice Hall International Editions
    Upper Saddle River , NJ - USA - 1998

- **And a lot of personal work -**